EARLY BIRD STORIES

Wild and Wacky Animals

An Alien's Guide

Hiya, I'm Zeek.

Alex Francis

Hi, I'm Finn.

Early ★ Reader

Lerner Publications ◆ Minneapolis

First American edition published in 2020 by Lerner Publishing Group, Inc.

An original concept by Alex Francis
Copyright © 2021 Alex Francis

Illustrated by Alex Francis

First published by Maverick Arts Publishing Limited

Maverick
arts publishing

Licensed Edition
Wild and Wacky Animals: An Alien's Guide

Lerner Publications Company
An imprint of Lerner Publishing Group, Inc.
241 First Avenue North
Minneapolis, MN 55401 USA

For reading levels and more information, look up this title at www.lernerbooks.com.

Main body text set in Mikado. Typeface provided by HVD Fonts.

Library of Congress Cataloging-in-Publication Data
The Cataloging-in-Publication Data for *Wild and Wacky Animals: An Alien's Guide* is on file
    at the Library of Congress.
ISBN 978-1-72841-507-9 (lib. bdg.)
ISBN 978-1-72841-539-0 (pbk)
ISBN 978-1-72841-540-6 (eb pdf)

Manufacured in the United States of America
1-48663-49086-1/21/2020

Table of Contents

Hello fellow aliens! What do you want to know about planet Earth?

INCOMING MESSAGE

Dear Finn and Zeek,

We want to visit Earth, but we're worried about some of the weird creatures on it!

Can you please show us some of the wackiest animals so that we are prepared?

From
Bim and Bam
(Planet Bland)

Over a very long time, animals have changed to help them live in their **environment.**

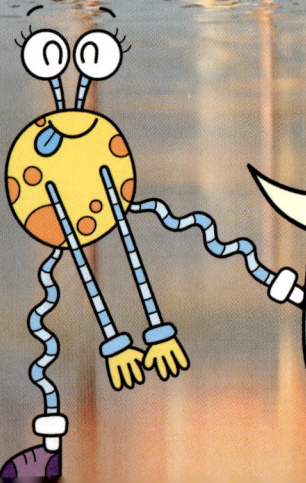

Flamingos have long necks and legs so they can go into deep water to get food.

This means there are some very wacky looking animals out there!

Also, flamingos are pink because of the algae and shrimp they eat!

Staying Alive!

Stick Insect

Some animals look like things around them so they aren't eaten. The stick insect looks like a stick. No animal wants to eat a stick!

Stick insects are usually green or brown, but some can be more colorful.

Stick insects mostly come out at night and eat leaves.

Flying Fish

Fish are best known for swimming, but these fish can also fly! They have winglike fins to help them take off. This means that they can fly out of the water, away from sneaky **predators**.

Fins

Is it a bird? Is it a plane? Nope, it's a fish!

Pufferfish

Before →

Pufferfish are very slow and clumsy swimmers. They puff up to make it difficult for predators to eat them.

After

Armadillo

Some animals have tough skin to protect them. The armadillo is covered in bony plates and has skin like leather. This **armor** is so tough, predators can't hurt the armadillo. Its name means little armored one in Spanish.

Some armadillos can even curl into a ball to avoid predators.

Fun and Funky!

Hoatzin Bird

The hoatzin bird is not just wacky because of its hairdo. It also smells really bad. Hoatzin birds are nicknamed stinkbirds! The smell like cow poop because of the plants they eat.

Phew!

They don't stink on purpose, but it helps them survive. Nothing wants to come near them!

Hoatzin chicks are born with claws on their wings, like dinosaurs!

Sea Hare

A sea hare's skin is covered in a **toxin** that makes it taste very bad. It can also squirt out an ink that smells horrible to other animals. Both these things stop predators from eating it!

Millipede

They smell even worse than my socks!

Although millipedes have lots of legs, they are very slow! To protect themselves, they give off a bad smell.

Luring Looks

Peacock

Sometimes animals use the way they look to attract other animals. Peafowl are one of these. They are big show-offs! Male peafowl are called peacocks.

They spread their tails out when trying to attract a female, called a peahen.

They have amazing,

long tail feathers.

Angler Fish

The angler fish is anything but beautiful! It lives deep in the ocean, where it is completely dark.

The female angler fish has a long **lure** on her head. This lure has a light on the end that attracts **prey.** An angler fish's mouth is so big, it can swallow prey twice its size.

Mammal Mix-Up!

Okapi

Some animals look like a weird mix of other animals.

Okapis are a little like giraffes and a little like zebras. They have long blue tongues so they can reach high leaves, like giraffes. They also have stripy legs like zebras to blend into their surroundings.

Okapi

Giraffe

Okapi

Zebra

The platypus is one of two **mammals** that lays eggs!

Platypus

Like the okapi, the platypus looks like a mix of other creatures.

The platypus has a bill and **webbed** feet like a duck.

It has a tail like a beaver.

It has a body and fur like an otter.

Dear Bim and Bam,

Animals on planet Earth are very wacky! But in most cases, they are wacky for a reason. Lots of them are just trying to avoid predators, so they won't bother you!

If you are visiting Mexico, perhaps you should go see this happy little creature called an axolotl.

From,
Finn and Zeek

26

Axolotl from Mexico

1. What do stick insects eat?
   a) Leaves
   b) Sticks
   c) Fish

2. What does *armadillo* mean in Spanish?
   a) Armored roller
   b) One with armor
   c) Little armored one

3. What is a female peafowl called?
   a) Peacock
   b) Peagirl
   c) Peahen

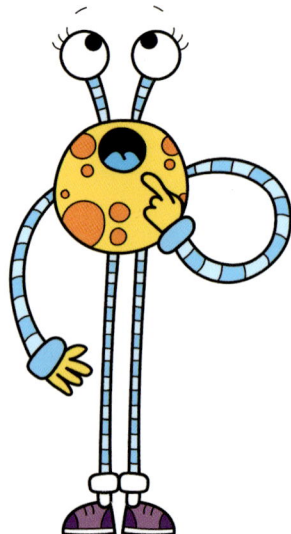

4. Where do angler fish live?
    a) In the deep, dark ocean
    b) In the shallow, light ocean
    c) In the dark underground

5. What color is the okapi's tongue?
    a) Yellow
    b) Blue
    c) Pink

6. The platypus is one of only two mammals that . . . ?
    a) Has fur
    b) Swims
    c) Lays eggs

Glossary

armor: something that is worn to protect the body from harm

environment: everything that surrounds a living thing

lure: something used to tempt an animal

mammals: warm-blooded animals with hair or fur that feed milk to their young

predators: animals that hunt other animals

prey: an animal that is hunted by another animal

toxin: a substance that is dangerous to living things

webbed: toes connected by skin

Index

Leveled for Guided Reading

Early Bird Stories have been edited and leveled by leading educational consultants to correspond with guided reading levels. The levels are assigned by taking into account the content, language style, layout, and phonics used in each book.

COLOR		GRL
Silver		L-P
Gold		K-L
Purple		J-K
Orange		H-J
Green		G-I
Blue		E-G
Yellow		C-E
Red		C-D
Pink		A-C

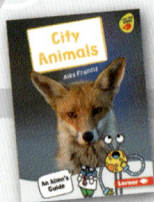
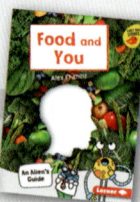
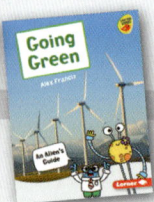
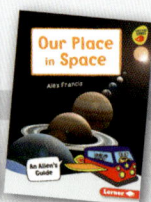
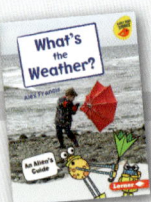
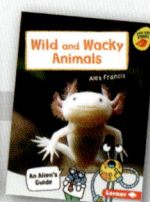